YOUR KNOWLEDGE HAS VALUE

- We will publish your bachelor's and master's thesis, essays and papers

- Your own eBook and book - sold worldwide in all relevant shops

- Earn money with each sale

Upload your text at www.GRIN.com and publish for free

Development of a predictive model for a velocity profile of a biofermenter using three different type of reactors

Chukwuemeka Ukpaka

Bibliographic information published by the German National Library:

The German National Library lists this publication in the National Bibliography; detailed bibliographic data are available on the Internet at http://dnb.dnb.de.

ISBN: 9783346328632
This book is also available as an ebook.

© GRIN Publishing GmbH
Nymphenburger Straße 86
80636 München

Print and binding: Books on Demand GmbH, Norderstedt, Germany
Printed on acid-free paper from responsible sources.

The present work has been carefully prepared. Nevertheless, authors and publishers do not incur liability for the correctness of information, notes, links and advice as well as any printing errors.

GRIN web shop: https://www.grin.com/document/977999

DEVELOPMENT OF PREDICTIVE MODEL FOR VELOCITY PROFILE OF BIOFERMENTER USING THREE DIFFERENT TYPE OF REACTORS.

Author: Ukpaka, C. Peter

Address: Department of Chemical/Petrochemical Engineering, Rivers State University, Port Harcourt

Abstract

This research contains the modeling of the velocity profile of a bioreactor with recycle; the concept of biochemical process. The biochemical process adopted is fermentation and a plug-flow fermenter (PFF) was taken as a case study. The derivation of workable model equations for monitoring and predicting the velocity profile of a PFF were obtained, together with obtaining the model equations for investigating the effect of microbial and substrate concentrations on the discharge coefficient, bioreactor's volume. Constant data were sourced from literatures, together with hypothetical values to simulate the derived model equations using Mathlab. Effect of biomass concentration on discharge coefficient, shows that increase in biomass concentration brings a corresponding increase in the discharge coefficient as well as the bioreactor's volume revealed that substrate concentration is depleting alongside with bioreactor's volume follows the same trend of change when substrate concentration is decreasing irrespective of whether the length or area of the bioreactor is varied. The effect of microbial concentration on bioreactor volume when area and length of bioreactor are varied reveals that the process followed same trend only that there is a presence of lag phase upon the influence of inhibitors.

1

TABLE OF CONTENT

1. INTRODUCTION 3
2. MATERIAL AND METHODS 5
   2.1   Derivation of model equations 5
3. RESULTS AND DISCUSSION 11
4. CONCLUSIONS 13
5. NOMENCLATURE 14
6. REFERENCES 15

1. INTRODUCTION

The variation in the concentration of substrate or microbial with respect to time. In other words, it is how fast or slow a biochemical reaction takes place. The velocity (reaction rate) profile of a bioreactor is monitored in any manufactured or engineered device or system that supports a biologically active environment.

For the purpose of this work we are taking a look at fermentation as a biochemical process taking place in a plug- flow Bioreactor (Fermenter) with recycle (Ukpaka, et al., 2009; Octave, 2007; William, 2007; Abashar and Butt, 200). Fermentation processes utilize microorganisms to convert solid or liquid substrates into various products. The substrates used vary widely, any material that supports microbial growth is a potential substrate. Similarly, fermentation-derived products show tremendous variety. Commonly consumed fermented products include bread, cheese, sausage, picked vegetables, beer, wine, citric acid, and soy sauce (Isla, et al., 1983., Jenning, 1991., Shah, 1969 & Copplestone).

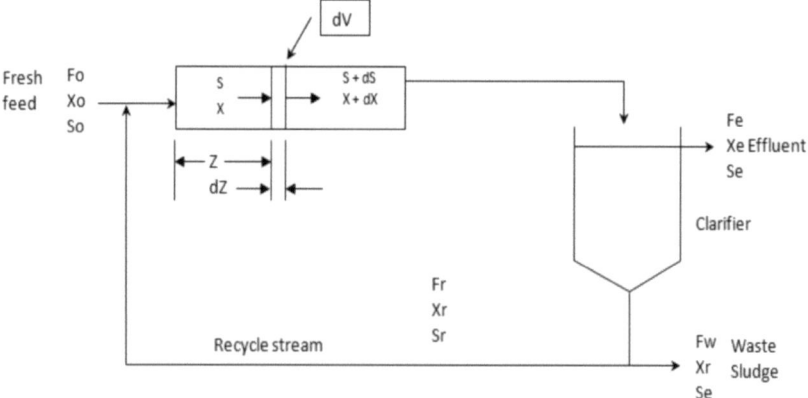

Fig. 1: Plug- Flow fermenter with recycle of biomass

In a bioreactor with recycle the effluent emerging from the reactor is fed into a settling unit. Microorganisms settle to the bottom of the tank, from where they are recycled into the reactor vessel. As a consequence of settling, the concentration of the microorganisms leaving the settling unit in the recycle stream is higher than that entering it from the biological reactor (singh and Saraf, 1981). The settling of the microorganisms greatly reduces their concentration in the effluent leaving the settling unit, producing a cleaner effluent stream. Recycle enables a higher concentration of microorganisms to be maintained in the bioreactor, which allows the reactor to run at much greater flow-rates and increase its efficiency (Ukpaka and Ogoni, 2015., Dyson and Simon, 1968; Hill, 1977; Elnashaie et al., 1988). The delimitation of Study includes: derivation of workable model equations to monitor the velocity profile of a plug-flow fermenter (PFF) with recycle and the use of MATLAB to simulate hypothetical values of a typical plug-flow fermenter using the derived workable model equations.

The velocity profile of a bioreactor is more or less known as the rate of reaction profile of a bioreactor. This defines the variation in the concentration of substrate or microbial with respect to time. In other words, it is how fast or slow a biochemical reaction takes place. This is measured by determining the change in the concentration with regard to time change. Enzymes are frequently used as catalysts to promote specific reactions in free solution. They are typically required in small amounts and are attractive in that they obviate both the need to provide the nutritional support which would be required for micro-organism to perform the same conversion, and the possible subsequent removal of those microbes. Furthermore, the enzyme need not necessarily be of microbial origin so that a wider choice of operating conditions and characteristics may be available.

2. MATERIAL AND METHODS

2.1 Derivation of model equations

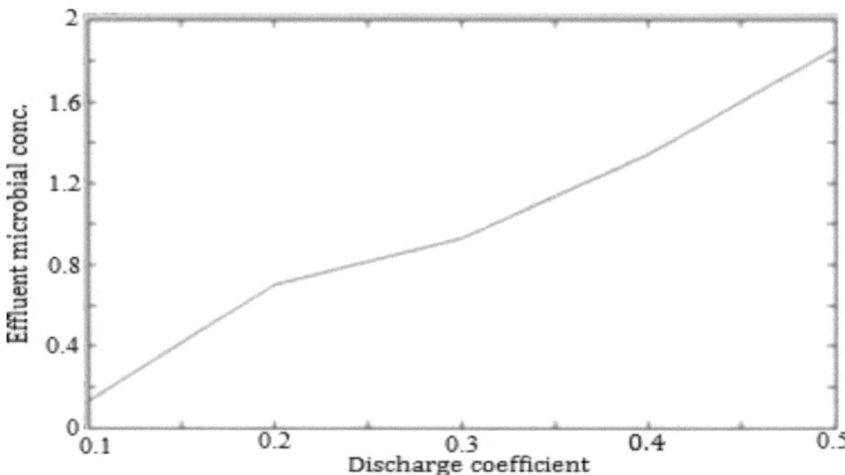

Fig. 2: Effluent microbial concentration (Xe) against discharge coefficient (Y)

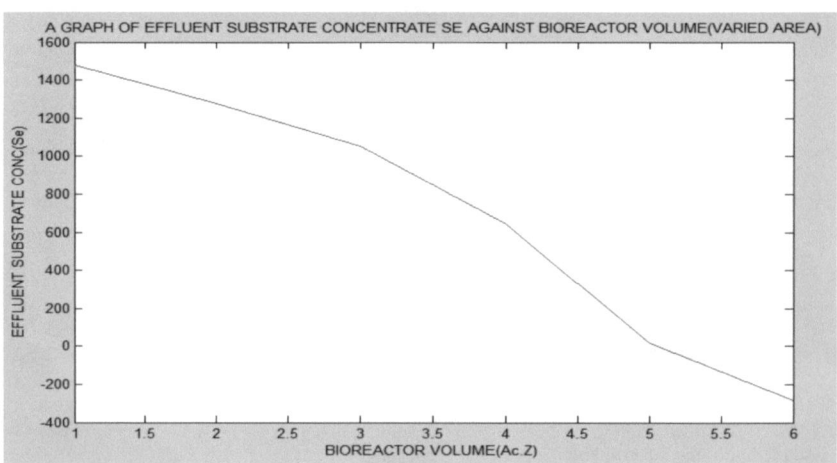

Fig. 3: Effluent substrate concentration Se against bioreactor's volume at varying bioreactor's cross sectional area ($a_c z$).

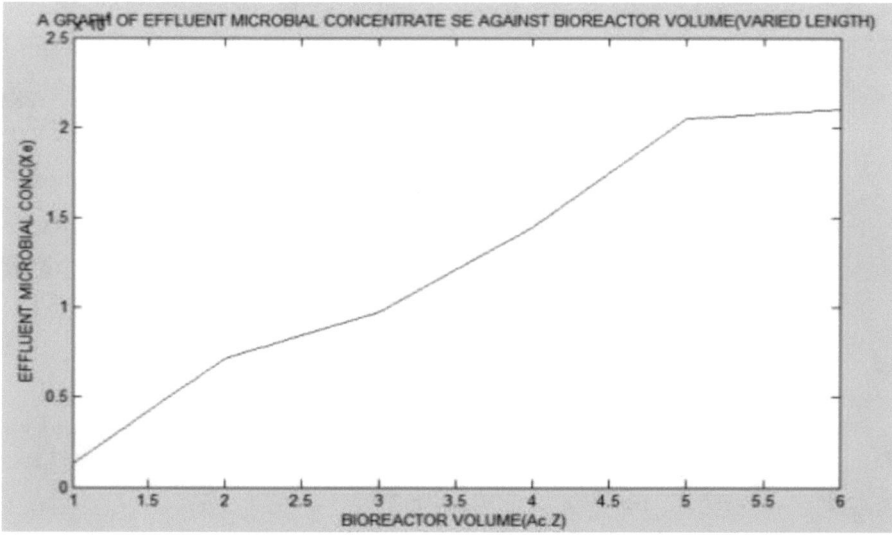

Fig. 4: Effluent microbial concentration (Xe) against bioreactor's volume at varying bioreactor's cross sectional area (a$_c$z). The curve represents that there is a corresponding increase in the bioreactor's volume as the population of the microbe's increases.

The growth of biomass is initiated by the addition of an inoculums; in this case it is represented by the stream at a volumetric flow rate of (Fo) with biomass concentration (Xo) and substrate concentration (So). As it exits the fermenter into the clarifier, a portion of it is recycled back into the fermenter and is mixed with the fresh feed as depicted above. It is assumed that there is no biochemical reaction or growth occurring in the clarifier, so that the substrate concentration S in the entering stream is the same as that in the clarified liquid effluent stream, in the recycle stream and in the exit biomass richstream.

The general material balance statement is;

Material input + Formation by Biochemical reaction - Material output=Accumulation (1)

Material Balance in terms of Biomass (X)

Rate of reaction R = Velocity V,

Where, input = $(F_O + F_R)X$

output = $(F_O + F_R)(X + dX)$

Formation by biochemical reaction = R_XdV, and

Accumulation = $\frac{dx}{dt}V$

$(F_O + F_R)X + R_XdV - (F_O + F_R)(X + dX) = (dX/dt)V$ (1a)

Considering when $dV = a_cdz$, and at steady state equation (1a)becomes;

$(F_O + F_R)X + RXa_cdz - (F_O + F_R)(X + dX)=0$ (1b)

Simplifying equation (1b) by substituting the above assumption we have

$RXacdz= F_O+F_R(dX)$ (1c)

Recycle ratio is given as $R = \frac{F_R}{F_o}$ and by substituting the expression into equation (1c), +

Therefore, $F_R = RF_O$

$R_Xacdz = F_O(1+R)dX$ (2)

Application of separating the variables into equation (2) and integrating, we have;

$$\frac{acdz}{F_o(1+R)} = \frac{dX}{dt}$$ (3)

Integrating equation (3) at the boundary conditions of 0-Z and X_A- X_e, we have;

$$\frac{ac}{F_o(1+R)}\int_0^z dZ = \int_{X_A}^{X_e} \frac{dX}{RZ}$$ (4)

7

Similarly, for the Substrate the form of the mathematical expression is given as:

$$\frac{ac}{F_0(1+R)} \int_0^z dZ = \int_{S_A}^{S_e} \frac{dS}{RS}$$ (5)

If, $S_A = \frac{S_0 F_0 + S_R F_R}{F_0 + F_R}$.,

$$S_A = \frac{S_0 F_0 + S_R F_0 R}{F_0 + F_0 R} \text{.,and} S_A = \frac{S_0 + RSR}{1+R}$$ (6)

Similarly, for the biomass we have

$$X_0 F_0 + X_R F_R = (F_0 + F_R) X_A$$

$$X_A = \frac{X_0 F_0 + XRFR}{F_0 + F_R}$$

If, $X_R = \mathcal{E} X_e$

$$X_A = \frac{X_0 F_0 + \mathcal{E} X_e F_0 R}{F_0 + F_0 R}$$

$$X_A = \frac{X_0 + \mathcal{E} X_e R}{1+R}$$ (7)

Substituting equations (6) and (7) into the boundary conditions on equations (5) and (4) as well as recalling equation (5) we have

$$\frac{ac}{F_0(1+R)} \int_0^z dZ = \int_{S_A}^{S_e} \frac{dS}{RS}$$

Recall that for Monod's kinetics and Discharge Coefficient, the reaction rate with respect to the substrate Rs may be defined as;

$$Rs = \frac{U_m S X_a}{(K_s + S)Y}$$

Substituting Rs into equation (5), we have

$$\frac{ac}{F_0(1+R)} \int_0^z dZ = \int_{S_e}^{S_A} \frac{(K_s + S)Y ds}{S X_a U_m}$$

$$\frac{ac}{F_0(1+R)} \int_0^z dZ = \frac{Y}{X_a U_m} \int_{S_e}^{S_A} \frac{(K_s + S) ds}{S}$$

8

$$\frac{acXaUm}{YF_0(1+R)}\int_0^z dZ = \int_{Se}^{SA}\frac{Ksds}{S} + \int_{Se}^{SA} ds \tag{7a}$$

Let $S_A = S_I$. Integrating the equation (7a), we have

$$\frac{acXaUmZ}{YF_0(1+R)} = Ksln\frac{Si}{Se} + (S_I + S_e) \tag{8}$$

The definition of the Discharge Coefficient $Y = \frac{Xe-Xi}{Si-Se}$ can be used to derive an expression for the

Biomass concentration by substituting $Y = \frac{Xe-Xi}{Si-Se}$ into equation (8) and expression obtained is given

below,

$$\frac{acXaUmZ(Si-Se)}{F_0(1+R)(Xe-Xi)} = Ksln\frac{Si}{Se} + (S_I + S_e)$$

$$\frac{acXaUmZ(Si-Se)}{F_0(1+R)} = (Xe - Xi)Ksln\frac{Si}{Se} + (S_I + S_e)(Xe - Xi)$$

$$\frac{acXaUmZ}{F_0(1+R)} = \frac{Xe-Xi}{Si-Se}Ksln\frac{Si}{Se} + (Xe + Xi)$$

$$\frac{acXaUmZ}{F_0(1+R)} = YKsln\frac{Si}{Se} + (Xe + Xi) \tag{9}$$

Let us recall that for a given substrate concentration S, the rate of biodegradation is expressed thus;

$$\frac{ds}{dt} = -\beta S \tag{9a}$$

On separation of variables for integration, the equation (9a) gives;

$$\int_{Si}^{Se}\frac{ds}{S} = -\int_0^t \beta dt$$

Let $t = \tau$

$$\ln\frac{Si}{Se} = \beta T \tag{10}$$

Substituting equation (10) into equation (8) gives;

$$\frac{acXaUmZ}{YF_0(1+R)} = Ks\beta T\frac{Si}{Se} + (S_I + S_e)$$

Assuming $Xa = Xi$

$$Se = \left(\frac{-acXiUmZ}{Y\,Fo(1+R)} + S_I \right) + (Ks\beta)\mathrm{T} \tag{11}$$

Equation (11) is the Model equation for predicting the velocity (reaction rate) profile in terms of substrate concentration of a plug-flow fermenter. Similarly, substituting equation (10) into equation (9) gives;

$$\frac{acXaUmZ}{F_o(1+R)} = YKs\beta\mathrm{T} + (Xe + Xi)$$

Assuming $Xa = Xi$

$$Xe = \left(\frac{-acXiUmZ}{Y\,Fo(1+R)} + X_I \right) - (YKs\beta)\mathrm{T} \tag{12}$$

Equation (12) is the Model equation for predicting the velocity (reaction rate) profile in terms of biomass concentration of a plug-flow fermenter.

3.   RESULTS AND DISCUSSION

For Fig 3 shown, which is the graph of effluent substrate concentration Se and effluent microbial concentration Xe against space time τ, the following equations where used to simulate the parameters;

$$Se = \left(\frac{-acXiUmZ}{Y\,Fo(1+R)} + S_I\right) + (Ks\beta)T$$

And $Xe = \left(\frac{-acXiUmZ}{Y\,Fo(1+R)} + X_I\right) - (YKs\beta)T$

The following parameters were chosen in simulating the developed model using Mathlab computer programme language, such as:

Fo = 18L/hr, Um = 0.25/hr, Ks = 0.12g/hr, Y = 0.42, R = 1, Z = 10m, ac = 0.5m2. T; 0, 5, 10, 15, 20, 25, 30 h, Influent Substrate: 2000, 1900, 1800, 1700, 1650, 1600, 1500, Influent Microbial: 2000, 1300, 6800, 9000, 13000, 18000, 18000, bioreactor's area: 0.1, 0.1, 0.2, 0.3, 0.4, 0.5, 0.6, Bioreactor's length: 2, 5, 7, 8, 10, 15, 20 and β = 1/hr

Fig 2 is showing the graph of effluent microbial Xe against discharge coefficient Y, the following data were used for simulation on this model equation;

$$Xe = \left(\frac{-acXiUmZ}{Y\,Fo(1+R)} + X_I\right) - (Ks\beta T)Y$$

The above graph shown in Figure 2 reveals that as the microbial population increases the discharge coefficient which defines the yield also increases. This shows that increase in microbial population have a positive effect on the yield.

Graph shown in Figure 3 reveals that decrease in substrate concentration leads to increase in bioreactor's volume. The curve justifies the fact that the smaller the concentration of substrate in a vessel the more the vessel appears to be large for such substrate, and if the substrate is increased the volume of the vessel will appear small for the substrate. For figure 3 showing the graph of

11

effluent substrate concentration Se against bioreactor's volume acz at varying bioreactor's length, the following data were used for simulating this model equation; $Se = \left(\frac{-XiUmZ}{Y\,Fo(1+R)}\right)a_cZ + (S_I + Ks\beta T)a_cZ$

Figure 4 is showing effluent microbial concentration Xe against bioreactor's volume acz at varying bioreactor's cross sectional area ac, the following data were used to simulate this equation;

$$Xe = (X_I - YKs\beta T) + (\frac{acXi}{Fo(1+R)})a_cZ$$

For Figure 5 showing effluent microbial concentration Xe against bioreactor's volume acz at varying bioreactor's length Z, the following data were used to simulate this equation;

$$Xe = (X_I - YKs\beta T) + (\frac{acXi}{Fo(1+R)})a_cZ$$

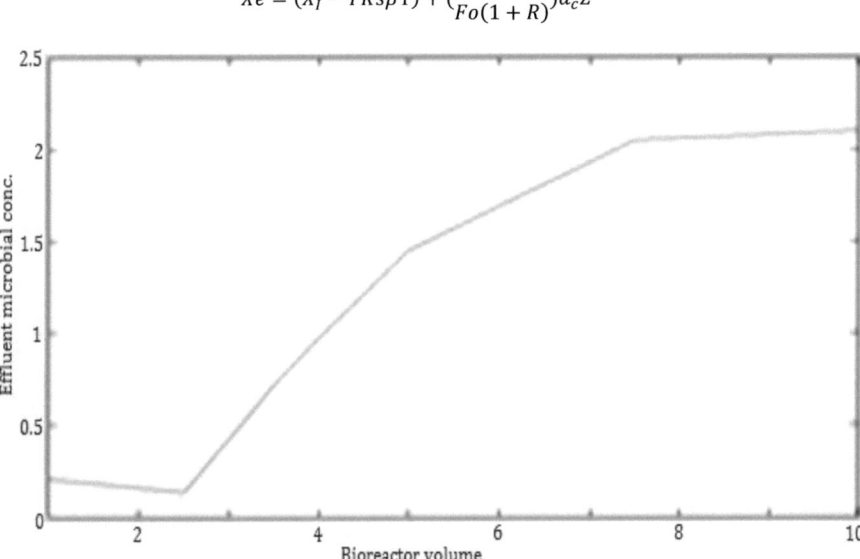

Fig. 5: Effluent microbial concentration xe against bioreactor's volume at varying bioreactor's length Z.

4.   CONCLUSIONS

The following conclusions were drawn from this research work. The Mathlab program was used for monitoring and predicting the rate of change in concentration at different space time. The relationship between the concentration and discharge coefficient was monitored using MathLab program. The relationship between the concentration and bioreactor volume was monitored using MathLab program. The mathematical approach of modeling the dynamics of a bioreactor with sludge recycle can be applied to other fields of engineering.

5.   NOMENCLATURE

ac = Cross Sectional Area of a Bioreactor

Fo = Volumetric Flowrate of Feed

Ks = Monod's Reaction Rate Constant

R = Recycle Fraction

Um = Maximum Specific Growth Rate

T = Space Time

Xi = Influent Microbial Concentration

Xe = Effluent Microbial Concentration

Xr = Recycled Microbial Concentration

Si = Influent Substrate Concentration

Se = Effluent Substrate Concentration

β = Reaction Rate Constant

Y = Discharge Coefficient

Z = Bioreactor's Length

6. REFERENCES

Abashar, M.E.E., (2003). Implementation of mathematical and computer modeling to investigate the characteristics of isothermal ammonia fluidized bed catalytic reactors. Mathematical and Computer Modeling 37(3-4), 439-456.

Butt, J. B., (2000). Reaction kinetics and reactor design: Taylor & Francis Publishers, 72-80.

Copplestone, J. C., Kirk, M. C., Death, S. L., Betteridge, N. G & Fellows, S. M. (2009) Ammonia, production, Springer published, 72.

Dyson, D.C., Simon, J.M., 1968. Kinetic Expression with Diffusion, Correction for Ammonia Synthesis on Industrial Catalyst. Industrial & Engineering Chemistry Fundamentals 7(4), 605-610.

Elnashaie, S.S., Abashar, M.E., Al-Ubaid, A.S., (1988). Simulation and optimization of an industrial ammonia reactor. Industrial & Engineering Chemistry Research 27(11), 2015-2022.

Hill, C.G., (1977). An Introduction to Chemical Engineering Kinetics and Reactor Design: John Wiley and Sons, 45-67.

Isla, M.A., Irazoqui, H.A., Genoud, C.M., 1993. Simulation of a ammonia synthesis reactor. 1. Thermodynamic framework. Industrial & Engineering Chemistry Research 32(11), 2662-2670.

Octave, L., (2007). Chemical Reaction Engineering, 3rd Edition, John Wiley and Sons India, 1000-1009.

Ukpaka C.P, Amadi S.A., Umesi, N., (2009). Modeling the physical properties of activated sludge biological wastewater treatment system in a plug flow reactor, The Nigeria Journal of Research and Production: A Multidisciplinary Journal 15, 134-147.

YOUR KNOWLEDGE HAS VALUE